This page intentionally left blank

WWW.JOURNALOFCOSMOINTEL.COM

Interuniversal Press

The Scientific Journal of Cosmointel
Vaughan, Canada

Journal of CosmoIntel

The First Scientific Journal in T-Consciousness Research

No. 1 | Apr. 2024
ISSN 2817-6995

Publication Details

Publication Title:
The Scientific Journal of CosmoIntel

Publisher: Interuniversal Press

Intellectual Property Rights:
Mohammad Ali Taheri

Managing Director and Editor-in-Chief:
Mohammad Ali Taheri

ISSN 2817-7002

Publication Language: Farsi - English

Year of Publication: 2024

Category: Scientific - Research

Website: WWW.JOURNALOFCOSMOINTEL.COM

Email: manager@journalofcosmointel.com

Interuniversal Press

Originator: Mohammad Ali Taheri

A Collection of Novel Theories in the Field of Cosmology by Mohammad Ali Taheri

Special Issue

T-Consciousness Cosmology

Inverse Cosmos

Important Point: A documentary concerning the topics addressed in this issue, titled:

“Inverse Cosmos”

was broadcast on

Jan 21, 2021

and is available for viewing on

the YouTube channel of Mohammad Ali Taheri

With special thanks to the research committee:

Sarah Hemmati
Mahyar Ramezanzadeh

WWW.JOURNALOFCOSMOINTEL.COM
WWW.COSMOINTEL.COM

Interuniversal Press

Preface

DOI: doi.org/10.61450/joci.v3iTC1EN.174

Introduction

Unveiling the Cosmos

The vast expanse of the cosmos, a celestial ocean teeming with constellations and swirling galaxies, has ignited an unquenchable spark of curiosity within humanity since the dawn of time. This eagerness to understand the universe in which we live has continually driven humankind to uncover the secrets of the cosmos. This endeavor is evident in ancient civilizations that charted constellations, as well as in modern astronomers who peer into the farthest reaches of space. Throughout history, pioneering astronomers like Galileo and Copernicus paved the way for a more tangible understanding of the universe. Eventually, modern cosmology, equipped with powerful telescopes and advanced detectors, has revolutionized our understanding of the universe as a whole, offering a framework for the origin, evolution, and ultimate fate of the universe.

From Philosophy to Physics Theories and Observations

The mysterious enigma of the cosmos has been and continues to be a pretense for the clash of ideas, inspiration, and countless arguments throughout human history. For tens of thousands of years, the configuration of stars in the sky has captivated our attention and inspired the creation of constellations from which myths and fictional stories were born, followed by philosophical inquiries essentially defining the philosophy of nature. Subsequently, the introduction of mathematics to simplify these philosophical concepts has opened up the world of physics to humanity. Persistent efforts and the accumulation of unanswered questions have not only compelled humans to observe and monitor the cosmos on larger scales with the advancement of technology but also aimed to interpret and analyze its components. This has been achieved through the processing of numerous mathematical models and the formulation of various theories, which sometimes align and sometimes contradict.

However, despite technical and technological advancements, the definition of the cosmos has been obscured in countless equations, calculations, models, and physics theories, with no clear and powerful unified interpretation provided thus far. Humanity remains at a loss in answering the questions: What is the cosmos? Why does it exist, and where is it headed? This is where the power of intuition and perception from a holistic viewpoint becomes essential for understanding and grasping the philosophy of the why and how of the cosmos. From this perspective, T-Consciousness Cosmology has attempted to unravel the complexities of conventional cosmology and provide a complete, clear, and unified interpretation of the cosmos through new theories. Additionally, this viewpoint addresses all concepts and the nature of what transpires in the cosmos, including Cosmic Information, Cosmic Mind, Cosmic T-Consciousness, and also various forms of Cosmic Life. These topics, however, do not entirely conform to the limited framework of currently accepted theories in physics and cosmology.

The Conventional Cosmological Perspective and Common Theories

Big Bang: The universe originated from an ultra-dense and hot point.

With the advancement of scientific research, the Big Bang theory was eventually established as the dominant cosmological model for the origin of the universe. This theory posits that the cosmos began nearly 13.8 billion years ago from an infinitesimal point with incredibly high density and temperature, known as the singularity. This singularity rapidly expanded and cooled in a fraction of a second, initiating the process of element formation and subsequently the formation of celestial bodies, ultimately leading to the cosmos we observe today.

From the perspective of cosmologists, the Big Bang theory is strongly supported by several key observations. One such observation is the discovery of the Cosmic Microwave Background (CMB). This radiation is considered a faint echo of the early universe, permeating the entire cosmos. In other words, its uniformity across the sky aligns with the predictions of a hot, dense origin of the universe. Additionally, the observed abundance of light elements such as hydrogen and helium in the universe also corresponds with the nuclear synthesis predicted to have occurred following the Big Bang.

Standard Model of Cosmology: A Framework for Cosmic Evolution

Building on the Big Bang theory, the standard model of cosmology (Lambda-CDM) provides another perspective on the universe's initial moments and its evolution, from the formation of fundamental particles to the creation of large structures like galaxies and galaxy clusters. This model incorporates the theory of inflation, a period of rapid exponential expansion believed to have occurred shortly after the Big Bang. The theory addresses the observed uniformity in the large-scale universe and theoretically resolves some of the problems with the Big Bang model. Additionally, the model relies on the existence of dark matter and dark energy, which are considered enigmatic components of the universe. Although invisible, dark matter influences the motion of galaxies and clusters through its gravitational pull. On the other hand, dark energy is believed to be responsible for the currently accelerated expansion of the universe.

Unanswered Questions and Ongoing Explorations

Despite the successes achieved in cosmology, the Big Bang theory and the standard model still face challenges. Questions about how the universe was born, the process of its evolution to its current form, the nature and geometric shape of the cosmos, its ultimate fate, the characteristics of dark matter and dark energy, and the possibility of other universes – along with countless unanswered questions – continue to drive ongoing research. Furthermore, alternative cosmological models such as the steady-state model, etc., are being examined to ensure a comprehensive understanding of the origin and evolution of the universe.

T-Consciousness Cosmology: A New Perspective on the Universe

Through its novel approach, T-Consciousness Cosmology comprises a collection of theories that examine and analyze topics such as the origin of the universe, its nature, the manner of its evolution, its fate, and hundreds of other theories.

As the name suggests, this viewpoint introduces a unique consciousness known as T-Consciousness. It posits that the universe, in addition to matter and energy, contains another element called T-Consciousness, which differs from definitions previously offered in the history of science or philosophy. From this perspective, it is argued that both matter and energy themselves arise from T-Consciousness.

Furthermore, T-Consciousness Cosmology articulates that the cosmos generally consists of two parts: frequency-based (~) and non-frequency-based (-):

The frequency-based part of the cosmos describes behavior that is periodic and non-linear, characterized by amplitude and wavelength, such that it has a non-continuous effect on the cosmos (i.e. all known types of waves and ordinary matter).

In contrast, the non-frequency-based part of the cosmos describes non-periodic and linear behavior, where the amplitude and wavelength are zero, and its effects in the cosmos are linear and continuous.

In this regard, for example, it can be said that space, gravity, and time themselves do not have a frequency effect and have a sustained effect on everything. Even if, for a moment, one of these, like gravity, were to exhibit a periodic effect, the entire cosmos would disintegrate. However, it is worth noting that the result of this influence is the emergence of particles (ordinary matter), which exhibit periodic and frequency-based behaviors. Similarly, if time itself were to have a periodic effect, the cosmos would likewise collapse in the same way, despite the fact that we have a periodic method of measurement for time (tick tock of a clock). Therefore, from this perspective, for the most part, the known physical aspect of the cosmos is periodic and frequency-based.

An important point to note is that the linearity of the impact of space, gravity, time, dark matter, and dark energy refers to the inherent influence of these factors in the universe, not the outcome of their effects.

Consequently, the frequency-based part (~) of the cosmos includes matter and energy, and the non-frequency, non-pulsing part (-) of the cosmos consists of two sections:

A- A section that in conventional cosmology is referred to with different definitions, such as spacetime, dark energy, and dark matter.

B- T-Consciousness, information, mind, life, dark life energy, etc., are parts that do not have specific definitions and are not mentioned in conventional cosmology. While from the viewpoint of T-Consciousness Cosmology, they constitute the main part of the cosmos.

Important Note: In T-Consciousness Cosmology, instead of the concept of "space-time," the term "space, gravity-time" is used, in which gravity and time are always proportionally intertwined and inseparable. In fact, the effect of gravity-time is considered as two sides of the same coin. Moreover, considering that if space did not exist, the cosmos would certainly not exist either. Therefore, from this perspective, space is considered a fundamental element of the cosmos, while it is neither matter nor energy. This means the nature of space, as one of the main components of the cosmos, is non-pulsing. This concept also applies to dark energy and dark matter, which this perspective identifies as functions of space itself.

Therefore, T-Consciousness Cosmology states that the structure of the components of the cosmos, such as dark energy and dark matter, is not composed of particles. Additionally, because of its non-pulsing nature, gravity is inherently a non-frequency element. Thus, generally, gravity is also not composed of particles (such as the hypothetical graviton particles in conventional science).

Regarding the non-pulsing nature of gravity or space, it can be noted that physics calculations show that celestial bodies with significant mass or acceleration can disturb spacetime in such a way that it appears as if gravitational waves propagate in all directions. In other words, conventional cosmology predicts that although they differ from each other, gravitational waves resulting from the spinning of neutron stars, the collision of black holes, and supernova explosions can be analyzed. However, T-Consciousness Cosmology defines what is commonly referred to as gravitational waves in physics simply as the squeezing and stretching of space due to the changing behavior of massive bodies in proximity to one another. Therefore, the changes in gravitational behavior caused by massive bodies only lead to the contraction and expansion of space. In simpler terms, gravity has a linear impact on the structure of space, not a wave-like one.

Like gravity, time exerts its influence in a linear fashion on the cosmos and its components, in tandem with gravity. If gravity were zero, time would also be zero. Conversely, if gravity approached infinity, time would similarly become infinite. It is also essential to mention that the type of timekeeping invented by humans (i.e. the ticking of a clock) is completely arbitrary, as time does not have a frequency or pulsing nature.

Consequently, T-Consciousness Cosmology uses "space, gravity-time" instead of the well-known term "spacetime."

The Origin and Fate of the Universe

Existing models in conventional cosmology have not yet provided a widely accepted theory about what existed before the Big Bang or how the various forms of matter and energy known today came into being at the initial moment of the explosion. This issue remains shrouded in ambiguity for cosmologists. In this context, T-Consciousness Cosmology, by introducing a new model named the 'Spherical Cosmos Model,' not only addresses the origin or how the initial seed of the cosmos came into existence, but also acknowledges the expansion of the cosmos and introduces a shell made of TAM (Taheri Absolute Matter), that isolates the cosmos. This model proposes a different foundational mechanism compared to conventional inflationary models and introduces a new concept called 'Space Rebound' to explain the increase in the volume of the cosmos. In fact, the theories of this viewpoint support each other in the understanding of the structure of the universe as a whole system, making simple predictions about the behavior of the cosmos. Moreover, T-Consciousness Cosmology, by addressing the nature of dark matter and dark energy and their functions, determines the cause of the cosmic expansion and its ultimate fate. Additionally, in line with the Spherical Cosmos Model (SCM), a new theory about another stage of the lifecycle of the cosmos, referred to as its 'Reversion,' is also proposed.

The Nature of the Building Blocks of the Cosmos

T-Consciousness Cosmology, in addition to addressing the general behavior of the cosmos, also explores the formation and function of its components, introducing new types of matter. According to the Spherical Cosmos Model, there is no contradiction between the formation mechanism of fundamental particles and the initial point of the cosmos (Big Bang). However, in the standard model of cosmology, which includes the theory of inflation and is based on general relativity and the standard model of fundamental particles, there is a clear contradiction known as the singularity at the birth of the universe and the formation of matter. Specifically, the singularity, a consequence of general relativity, is an obstacle that is inconsistent with the formation of fundamental particles in the initial moments of the cosmos's birth.

Overall, it can be stated that T-Consciousness Cosmology offers a unique view of the cosmos by altering the perspective of the observer. From this shifted viewpoint, the cosmos is perceived as a grand system endowed with distinct identity, personality, and behavior. This system not only follows a specific trajectory to fulfill a special purpose but also demonstrates a high level of intelligence.

The Multiverse from a New Perspective

T-Consciousness Cosmology asserts that the cosmos in which we currently live follows a sequential principle (Consecutive Cosmos) and has its own lifecycle. It is one of countless homogeneous or heterogeneous universes, each with its own unique characteristics and behaviors (laws of physics).

Additionally, from this viewpoint, the fundamental constants of physics change according to different cosmic epochs and locations. For example, gravity-time will range from infinity at the beginning of the cosmic lifecycle to zero at the terminal edge of the cosmos (the ultimate stage of space rebound).

and…

Originator of T-Consciousness Cosmology: Mohammad Ali Taheri

The Inverse Cosmos

DOI: doi.org/10.61450/joci.v3iTC1EN.175

Abstract

The Big Bang Model is the most widely accepted cosmological model regarding the origin of the universe, resting on two fundamental assumptions: one is Einstein's theory of general relativity, describing the gravitational interaction of matter with space-time, and the other is the cosmological principle, stating that an observer viewing the universe will be faced with homogeneity and isotropy, meaning that the universe has no edge. According to this model, one cannot seek a specific point of origin of the big bang, as this explosion occurred simultaneously throughout the whole universe. Through the evolution of various cosmological models, the Standard Cosmological Model (Lambda-CDM) also assumes that the universe consists of three main components: 1. A cosmological constant associated with dark energy, 2. Cold dark matter, and 3. Ordinary matter. From the perspective of cosmologists, this model provides a general explanation for a wide range of observed phenomena, including the abundance of light elements, cosmic microwave background radiation, and large-scale structures in the universe. However, there still exist aspects of the observed universe that currently pose challenges not explained by the big bang model. In response to these challenges, "*T-Consciousness Cosmology*" proposes multiple hypotheses, offering not only a new interpretative angle on the behavior and function of the components of the cosmos, but also stating that the latest cosmological models have paradoxes relative to the observations made. T-Consciousness Cosmology asserts that if we accept the prevalent models of cosmology, we are then faced with illogical manifestations of an expanding universe. In this discussion, several reasons are presented under the concept of an "*Inverse Cosmos*," challenging certain interpretations of the standard cosmological model or the latest big bang model regarding the origin of the cosmos, its geometric shape, and the stages accepted by most cosmologists for cosmic expansion. These challenges include issues such as the *Density Obscurity Horizon* for *Matter* and *Energy*, the *Thermal Obscurity Horizon*, the *Timelessness Horizon*, and others.

Keywords: T-Consciousness Cosmology, Inverse Cosmos, Density Obscurity Horizon for Matter and Energy, Thermal Obscurity Horizon, Timelessness Horizon

Conventional Cosmology

Preface

Since ancient times, humans believed that the Earth was at the center of the universe. However, it was later understood that the Earth and other planets of the solar system all revolve around the Sun. This led to the belief that the Sun was at the center of the universe. This belief was soon abandoned as well, as it became clear that stars are not uniformly distributed in the sky; instead, their disc-like distribution forms the Milky Way galaxy. Consequently, it was realized that the Sun, which had been considered the center of the universe, is actually located at a distance of two-thirds the radius of the Milky Way galaxy from its center.

Humanity then endeavored to place our galaxy at the center of the universe. However, observations of the depths of space revealed the reality that the Milky Way is just one of billions of galaxies in our current cosmos.

Principles of Modern Cosmology

As observational equipment advanced, cosmologists concluded that, regardless of where we are in the universe, the universe appears uniform; in other words, the universe is homogeneous and isotropic (Figure 1). In cosmology, homogeneity means that observers have access to the same observational evidence from different locations; however, this evidence includes only the part of the universe we can see. Isotropy means that the universe, statistically, appears the same in all directions. Therefore, homogeneity and isotropy, which are fundamental principles of modern-day cosmology and form the theoretical basis of the big bang theory regarding the observable evolution of the cosmos, state that the cosmos has no preferred location or direction.[1 2 3]

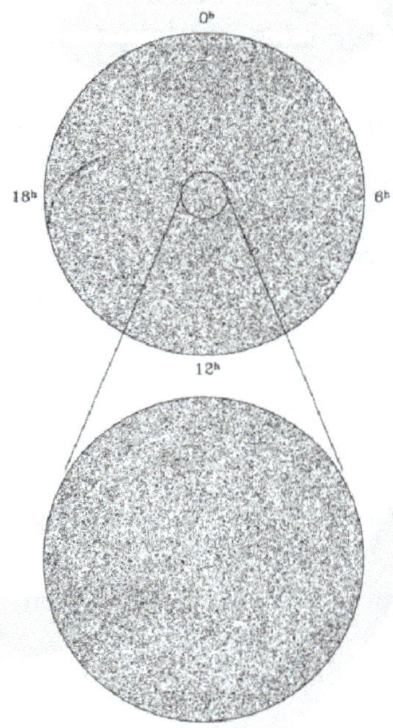

Figure 1: Large-Scale Universe: The points in the upper figure represent the positions of the brightest radio sources observable from the northern hemisphere, while the lower figure displays a comparable number of fainter sources within 15 degrees of the North Pole. The isotropy of these two figures, as seen in the sky, confirms that the universe is spatially homogeneous on the largest scales.

Reprinted with permission from [Ref: doi: 10.1073. pnas.96.9.4756], Copyright (1999) National Academy of Sciences, U.S.A

An important point to note is that cosmological principles are only applicable at large cosmic scales and do not hold at smaller scales (Figure 2).

Figure 2: Inconsistency of the homogeneity and isotropy principle in the small-scale universe

The following example is provided for a simpler description of the concepts in figures 1 and 2. Such that if an observer looks at a glass of milk, they see it as a completely homogeneous and uniform liquid; however, if the same observer examines the milk at a smaller scale, they encounter a non-homogeneous and non-uniform world (Figure 3).

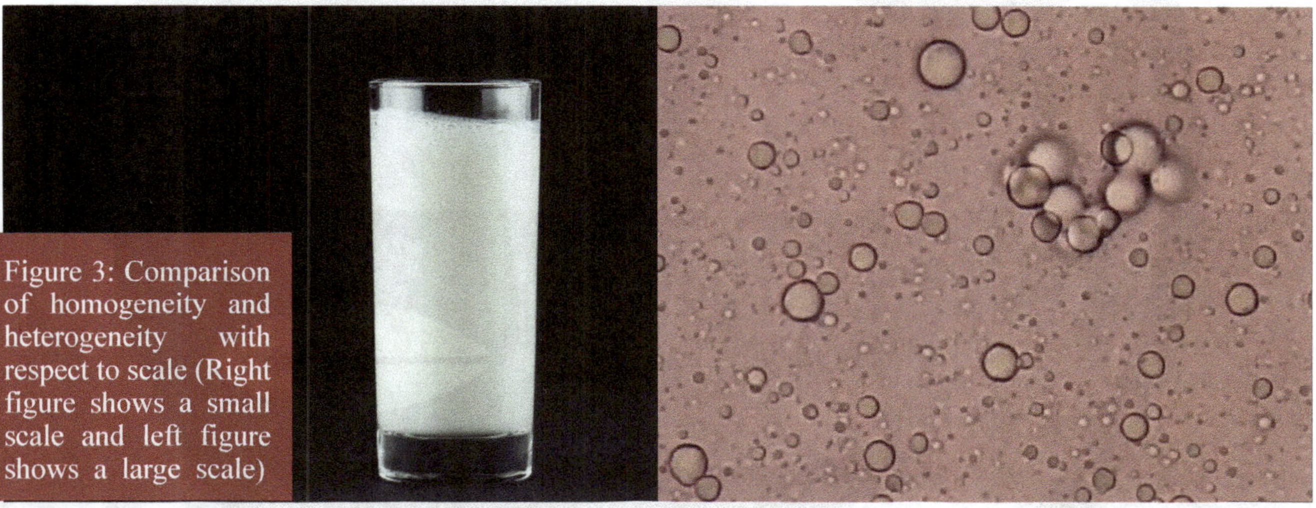

Figure 3: Comparison of homogeneity and heterogeneity with respect to scale (Right figure shows a small scale and left figure shows a large scale)

On large cosmic scales, this concept also holds true; in the context of conventional cosmology, when we consider the distribution and velocity of millions of galaxies in the vast universe approximately 13.84 billion years old, we conclude that the universe is homogeneous and uniform everywhere. However, this is not the case when we observe the universe on smaller scales.

Therefore, cosmological principles with the characteristic of the "large-scale universe" refer to a universe beyond the observable local galaxies.

On the other hand, observations and studies conducted so far, including the discovery and examination of the Cosmic Microwave Background (CMB), indicate that the temperature and density of matter at the time of the emission of these rays were uniform and homogeneous in all directions in the depths of the universe (Figure 4). From the perspective of conventional cosmology, when we look at the early universe through the lens of this radiation, we find that the distribution of matter and energy in the universe lacks any special characteristics; meaning that there are no distinct structures to indicate what transformation occurred between the initial big bang and the present time.[1]

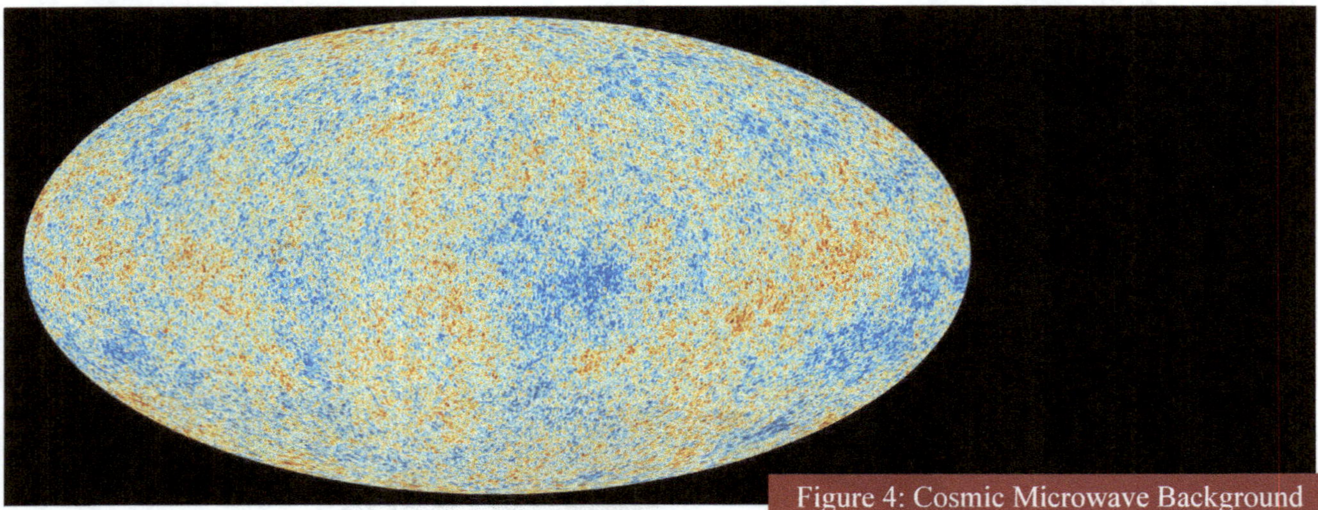

Figure 4: Cosmic Microwave Background

Cosmological evidence indicates that the cosmos has been expanding since the occurrence of the big bang. As observed in astronomical data, all celestial bodies are moving away from us. An important point to note is that the greater the distance of a celestial body, the faster it is moving away from us. These velocities are measured using a process known as "redshift" (Figure 5). The cosmic redshift in the standard FLRW metric is commonly explained as the elongation of a photon's wavelength due to the expansion of space.[4 5 6 7] The Friedmann–Lemaître–Robertson–Walker metric is based on the precise solution of Einstein's general relativity field equations. It describes a homogeneous and isotropic universe that is expanding. Simply put, redshift means the shift of the spectral lines of celestial objects toward longer wavelengths or red, which occurs due to these objects moving away from us.[7]

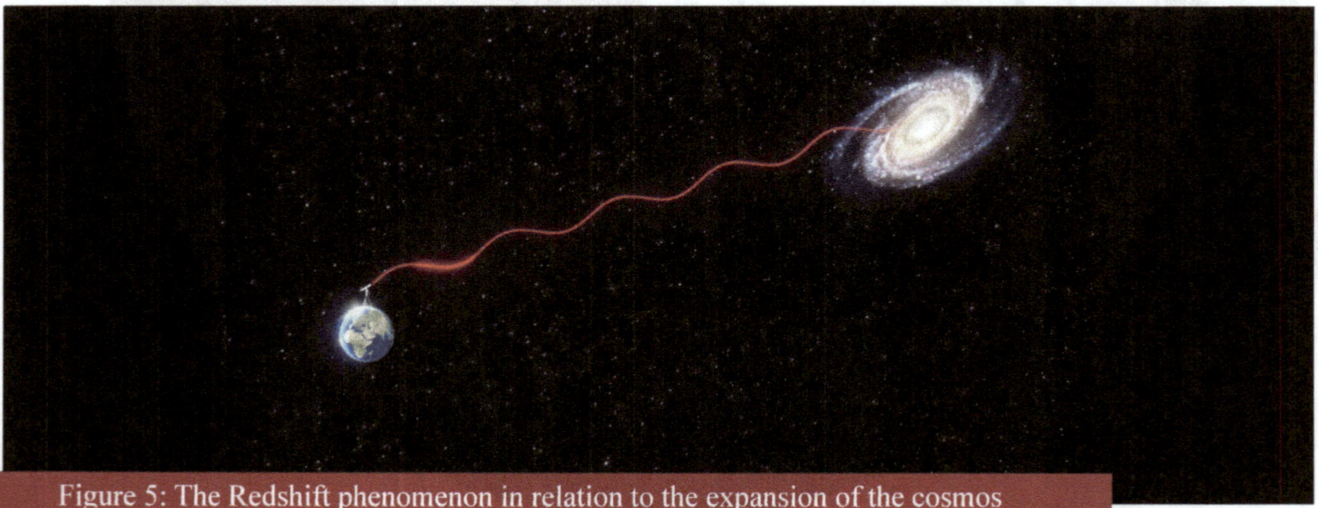

Figure 5: The Redshift phenomenon in relation to the expansion of the cosmos

The expansion of the cosmos is a key feature in cosmology related to the big bang phenomenon and signifies the increasing metric distance between celestial bodies over time. This can be visualized by imagining an uninflated balloon. If we draw dots on the balloon and then inflate it, we observe that as the volume of the balloon increases, the distance between the dots on the balloon also increases. This serves as an analogy for the concept of an expanding cosmos (Figure 6). However, an important point to note is that, from the perspective of conventional cosmology, the expansion of the universe is internal and relates to the relative distances between objects. It does not account for the stretching of space between objects, nor do these objects move toward space outside of the cosmos.[8 9 10 11 12]

Figure 6: Cosmic expansion as a metric distance increase between celestial bodies

The big bang theory is an accepted cosmological model that explains how the cosmos expanded from an initial state of very high density and temperature. It does not describe the cause of the cosmos's initial explosion; rather, it investigates the consequences of this grand explosion. According to this theory, the cosmos was concentrated at a very small point about 13.8 billion years ago. This initial state has since evolved into the vast and much colder cosmos we currently inhabit. The theory also faces challenges, such as the horizon problem and the flatness of the universe. In response, a theory called "inflation" describes a phenomenon known as exponential expansion, addressing these challenges and outlines the mechanism of the explosion in the early moments. Inflation, despite some opposition, is currently accepted by most cosmologists in modern cosmology. It suggests that in the early moments of the big bang, specifically in the first trillionth of a second, the cosmos expanded by a factor of approximately ten to the power of fifty (Figure 7). In fact, inflation predicts that primordial gravitational waves in the cosmos, like early density fluctuations (baryon acoustic oscillations), are produced by quantum fluctuations.[13]

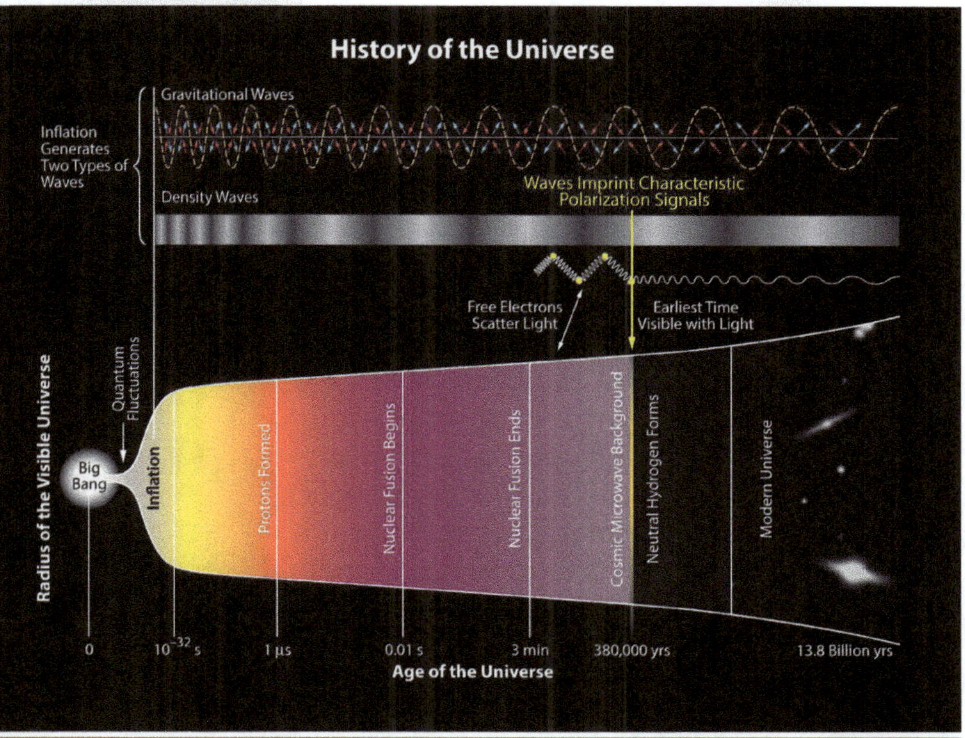

Figure 7: The timeline of the universe's expansion in the big bang model from the past to present

Credits:
BICEP2 collaboration
CERN/NASA

Although the big bang theory has provided humanity with information about the cosmos, it has also left many unanswered questions for cosmologists, with new mysteries being added every day. Despite observational data and their interpretations, increasingly complex paradoxes continue to emerge. Therefore, it seems that to analyze observational data and understand the mechanisms of such a vast cosmos, we need to study the structure of the cosmos and the behavior of its components from a new perspective.

T-Consciousness Cosmology

By proposing the Spherical Cosmos Model, among the hypotheses proposed by Mohammad Ali Taheri, T-Consciousness Cosmology offers a novel interpretative angle to some observational findings. These findings have been accepted in the latest big bang model but their analysis has been challenging. It points out the paradoxes between observational findings and the interpretations provided by conventional cosmology, presenting these paradoxes under the topic of the "Inverse Cosmos." Essentially, this perspective suggests that if we accept certain aspects of the latest big bang model or the Standard Cosmological Model, we are confronted with the inverse manifestations of the cosmos.

Reason 1:

A significant point raised on the topic of the Inverse Cosmos is that, according to the big bang theory, the cosmos initially condensed to a very dense point. After the explosion and subsequent expansion, it is logical that the density of matter should decrease as the volume of the cosmos increases. This means that, as we look into the depths of space around us, we should see the universe as more rarified in areas farther from the point of the explosion, where the spread of matter has expanded due to spatial expansion. However, cosmologists currently observe a uniform density of matter and energy in all directions of the cosmos up to a depth of approximately 13.84 billion light-years. This uniformity contradicts the logical expectation of a decrease in density with distance from the center of the big bang, as per the model in conventional cosmology (Figure 9).

In other words, the observed uniform density of matter in the cosmos, accepted through the principle of homogeneity and isotropy in conventional cosmology, contradicts the expectation of a more rarified universe over time with increasing volume. Cosmologists have upgraded the big bang model and, using the theory of inflation, not only explain the isotropy of the Cosmic Microwave Background (CMB) but also address issues such as the flatness problem and the horizon problem, including magnetic monopoles. It is worth mentioning that this is a theoretical hypothesis and, despite its acceptance in the cosmology community, still has its detractors[14]; given this, T-Consciousness Cosmology raises the challenge that:

The absence of observed decrease in matter, density, or celestial bodies in the expanding space, resulting from the initial explosion, indicates the inverse manifestations and behavior of the cosmos, and the inflation theory cannot adequately address this issue (Figure 8).

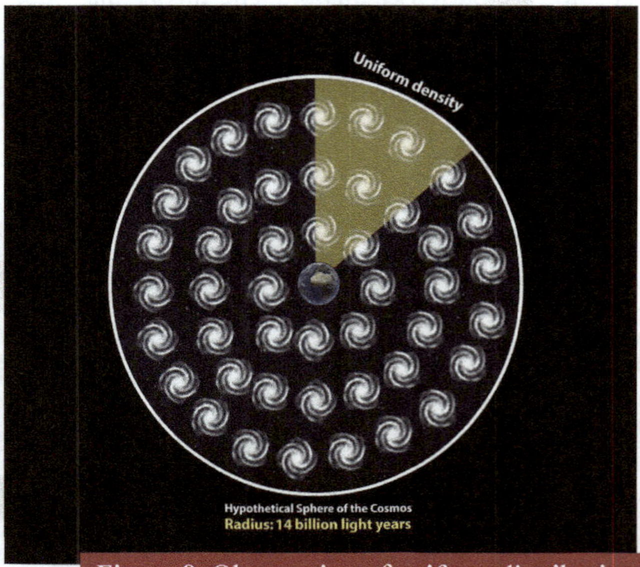

Figure 8: Observation of uniform distribution and density of matter and energy in all directions of the cosmos

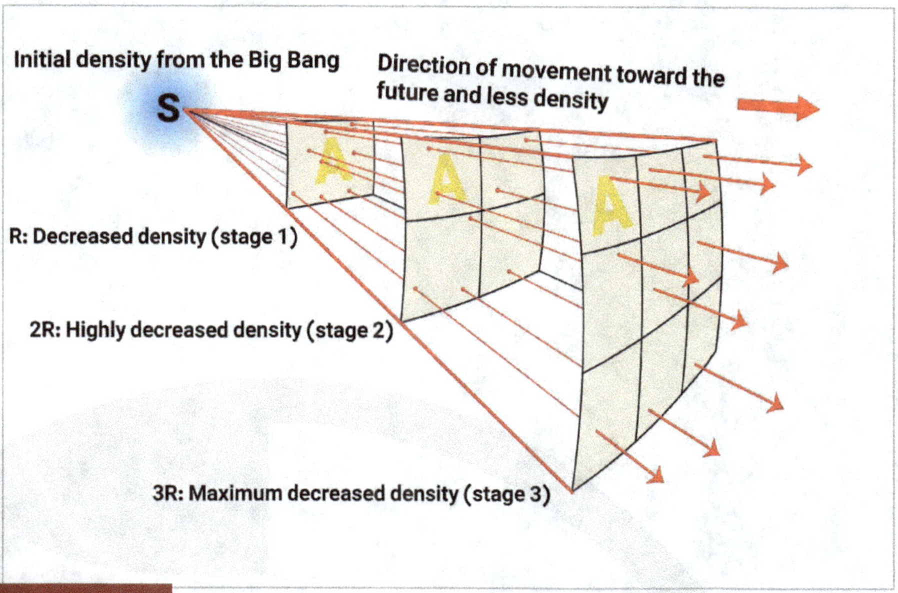

Figure 9: In this figure, S represents the surface and R represents the radius

The requirement for a logical decrease in density as we move further from the center of the Big Bang

Reason 2:

Another notable point is that if, according to reason one, cosmologists look in any direction, the prevailing cosmological view which states the two important characteristics of homogeneity and isotropy of the cosmos are still preserved, implies they will not observe the future of the cosmos due to this uniformity that exists in all directions on a massive scale. On the other hand, the ultimate fate of the cosmos in the standard cosmological model depends on factors such as its overall shape, the amount of dark energy, and ultimately on an equation that determines how dark energy density responds to the expansion of the cosmos.[15 16 17] In this context, cosmologists envision three possible fates for the cosmos: one is the Big Rip, where dark energy overwhelms matter density. Another is the Flat Universe, where the universe continues to expand indefinitely, but this expansion will slow down and eventually stop when time reaches infinity. The last one is the Big Crunch, where matter density overcomes dark energy, and the universe collapses into itself due to gravity.[17]

Therefore, in the accepted big bang model, the depths of space represent the past of the cosmos. Now, from the perspective of T-Consciousness Cosmology, the issue raised is that if an observer on Earth looks into the depths of the cosmos, they cannot observe galaxies that indicate the progression of the cosmos's expansion into the future and are only faced with the past of the cosmos; they observe galaxies that are approximately close to the age of our galaxy or younger galaxies. Thus, it can be concluded:

From the vantage point of an observer on Earth, the uniformity of matter and energy distribution in all directions makes it impossible to envisage a fate for the cosmos based on observations and only allows for conclusions based on probabilities. This is another reason for the inverse behavior of the cosmos (Figure 10).

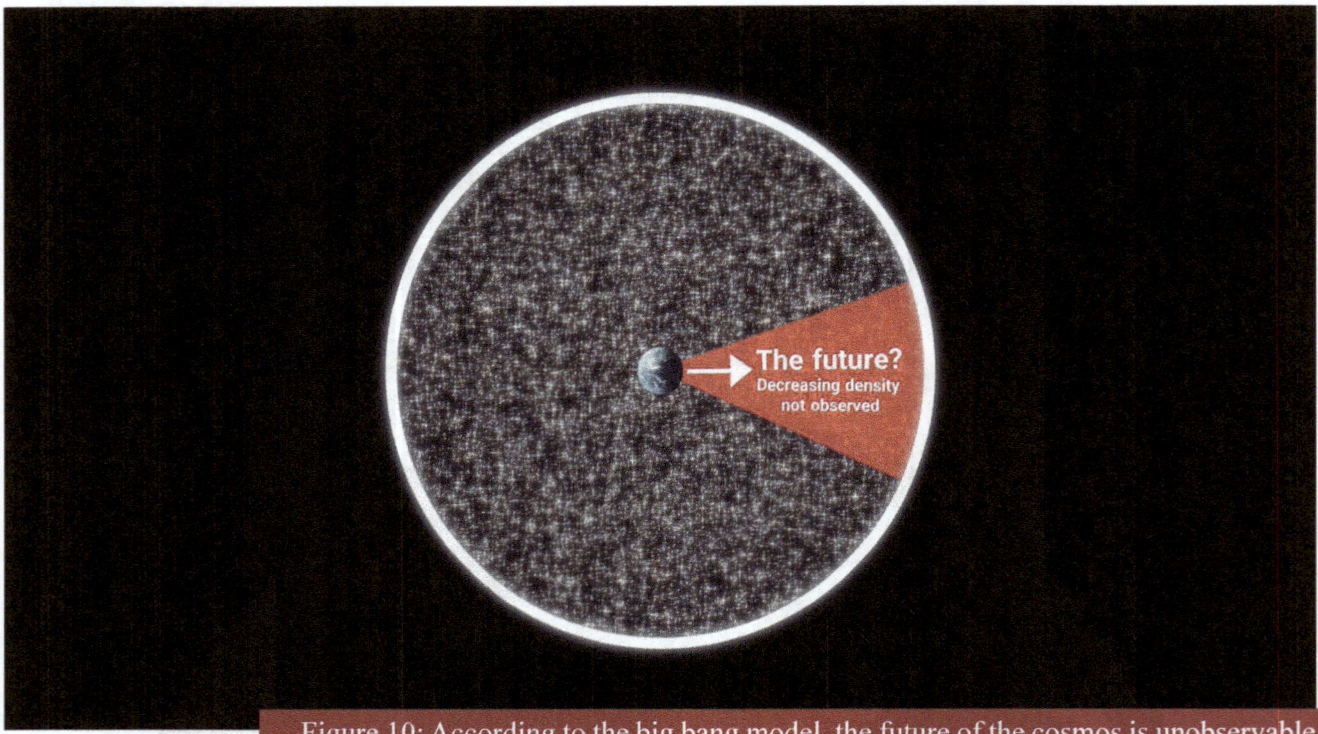

Figure 10: According to the big bang model, the future of the cosmos is unobservable due to the uniform distribution of objects

Reason 3:

According to the big bang theory, as the cosmos expands, our observational horizon in the depths of space is continually increasing. Therefore, from the perspective of conventional cosmology, when we look at the edge of our observational horizon, approximately fourteen billion light-years deep, we not only observe signs of the big bang, one of which is the Cosmic Microwave Background (CMB) that fills the entire cosmos, but also encounter a higher density of objects like mature galaxies in the depths of space in the process of inflation. Moreover, if we change our viewing angle in any direction of this celestial sphere, we still face the same increasing density and radiation indicative of the cosmic origins in the time direction pointing towards the past. In essence, if we look in any direction of the cosmic depths with advanced telescopes like James Webb, we encounter massive clusters of galaxies that are not only increasing but also encompassing us (Figure 11). It is noteworthy to mention that in conventional cosmology, the initial point of the cosmos no longer exists due to expansion and only its effects, such as the CMB, are observable and analyzable.[18 19]

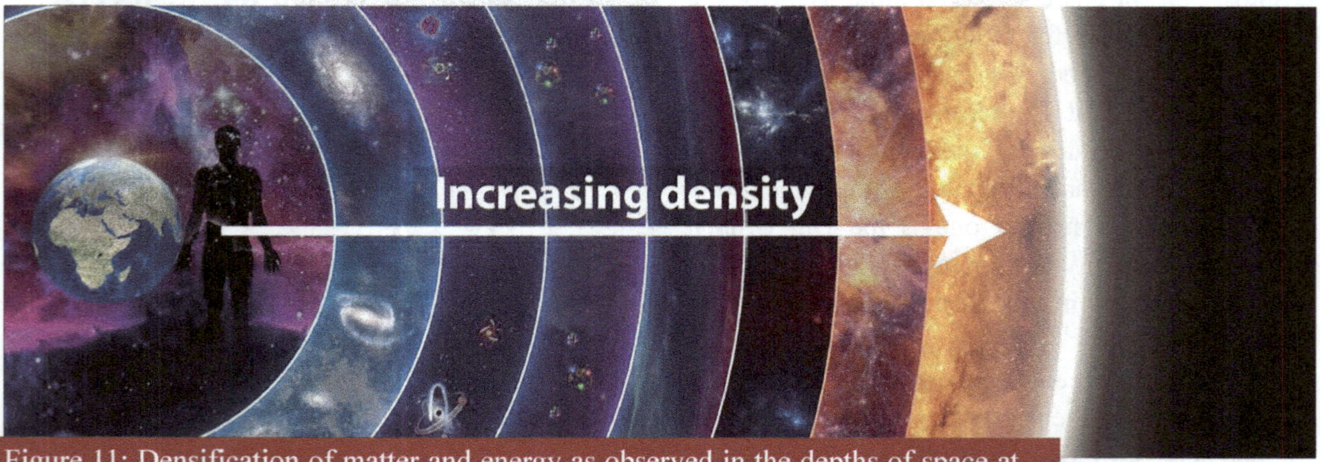

Figure 11: Densification of matter and energy as observed in the depths of space at the horizon of sight

Considering this, T-Consciousness Cosmology states that if we compare the current density of matter and energy in the cosmos with the past density, we find that the ratio of this density in the vast expanding space today is increasing as the surface of space in the universe increases. Thus, the cosmos operates inversely; where we expect the expansion of the cosmos to lead to a dilution of this density, in our observational horizon, matter and energy become more dense. Consequently, from the perspective of T-Consciousness Cosmology:

From any direction we look at the depths of space, according to the big bang model, we still reach the beginning point of the cosmos, which we are surrounded by, and in this observational horizon, it appears that the density of objects in the depths is increasing with every moment of the cosmos's expansion (Figure 12). Therefore, it can be concluded that the interpretation that we observe the early stages of the cosmos's expansion by observing the depths of space is unjustifiable, and here also, the cosmos operates in an inverse manner.

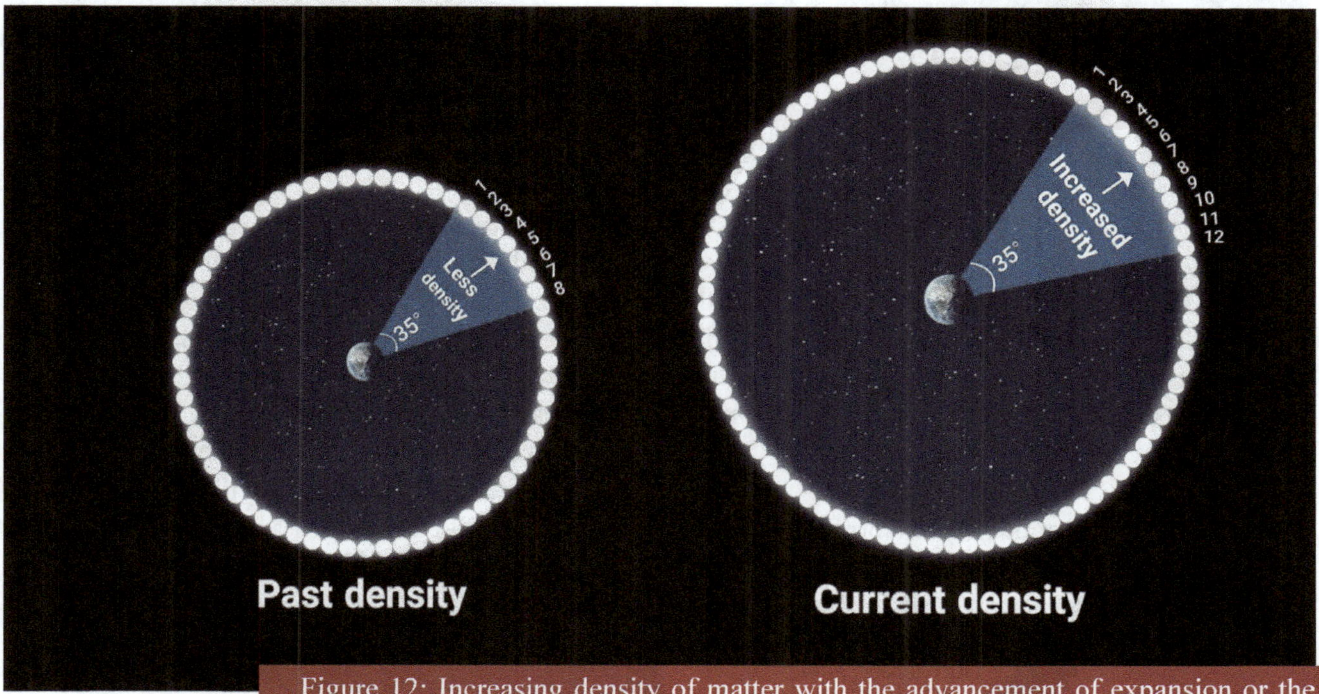

Figure 12: Increasing density of matter with the advancement of expansion or the increase in the volume of the cosmos

T-Consciousness Cosmology, considering this reason, states that if we accept this interpretation, in the big bang model, the direction of decreasing density of matter and energy vector, which is drawn from Earth in 360 degrees toward every direction in the depths of space (R_2), merges hypothetically with the direction of the vector representing the decrease in density of matter and energy drawn from the edge of our observational horizon, i.e., from a distance of approximately 13.84 billion light-years toward us (R_1). This integration results in the actual direction of the cosmos's expansion becoming indistinct at a certain distance from Earth. This leads to the formation of a hypothetical spherical horizon with a radius of R_2, which this perspective terms as the ***"Density Obscurity Horizon for Matter and Energy Density."***

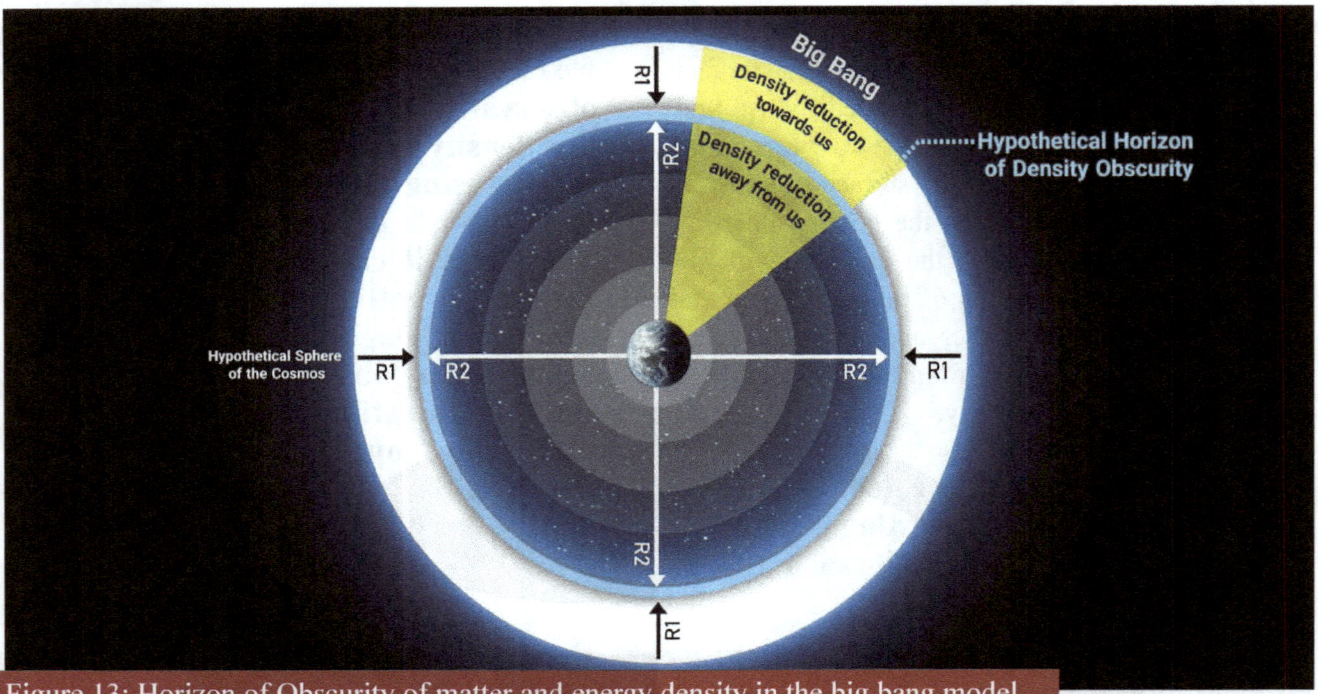

Figure 13: Horizon of Obscurity of matter and energy density in the big bang model from the perspective of T-Consciousness Cosmology

Reason 4:

The flat, Big Rip, and Big Crunch models and the fates predicted for each of them, based on factors like critical density, have not only been challenged by new findings but also remain theoretical and unconfirmed in the realm of conventional cosmology.[14 16 17]

Therefore, if these models are correct, from the perspective of T-Consciousness Cosmology, one of the peculiar issues that arise is that while the cosmos has maintained its density of matter and energy, it appears as if this density has also been magnified simultaneously! In other words, contrary to our expectation that objects in the depths of the cosmos should appear more rarified, their density, from the perspective of an observer on Earth, has significantly increased (Figure 14). The extent of this magnification can be calculated by comparing the surface area of the current hypothetical sphere of the cosmos with a radius of approximately fourteen billion light-years to the surface area of the Cosmic sphere fourteen billion years ago:

$$\text{Current Magnification Value} = \frac{\text{Surface area of the hypothetical cosmic sphere fourteen billion years ago}}{\text{Current surface area of the hypothetical cosmic sphere with a radius of fourteen billion light-years}}$$

$$\frac{4\pi \times 14 \times 14 \times 10^{48}}{4\pi \times 10^{-100}} = 1.96 \times 10^{150}$$

This calculation shows that the current magnification ratio is 1.96×10^{150} times the magnitude at the initial moment of the big bang.

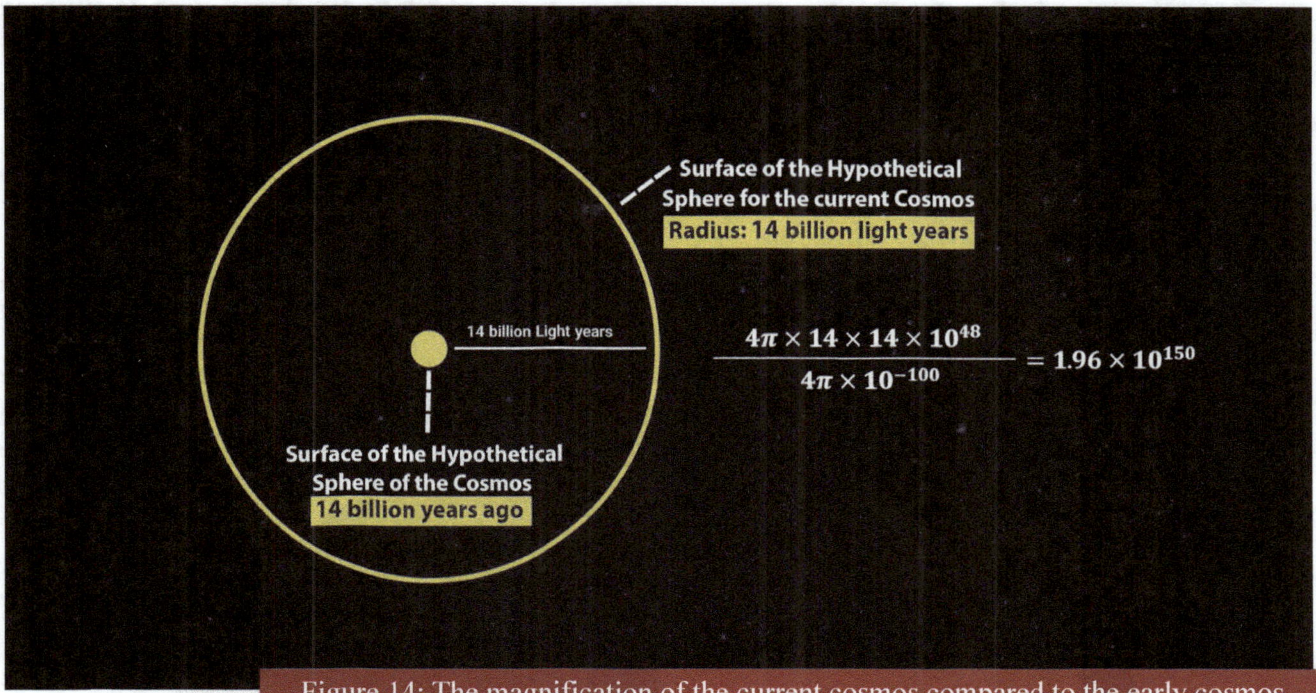

Figure 14: The magnification of the current cosmos compared to the early cosmos

It can also be calculated that after another two billion years, the magnification of the cosmos compared to its beginning at the big bang will be several times the current level. This amount is equal to the ratio of the surface area of a hypothetical sphere of the cosmos with a radius of sixteen billion light-years to the surface area of a hypothetical sphere fourteen billion years ago. That is:

$$\frac{4\pi \times 16 \times 16 \times 10^{48}}{4\pi \times 10^{-100}} = 2.56 \times 10^{150}$$

Which shows a 2.56×10^{150} times magnification.
Or in accordance with figure 15, the magnification of the cosmos in 2 billion years relative to the current cosmos is equal to:

$$\frac{4\pi \times 16 \times 16 \times 10^{18}}{4\pi \times 14 \times 14 \times 10^{18}} = 1.3$$

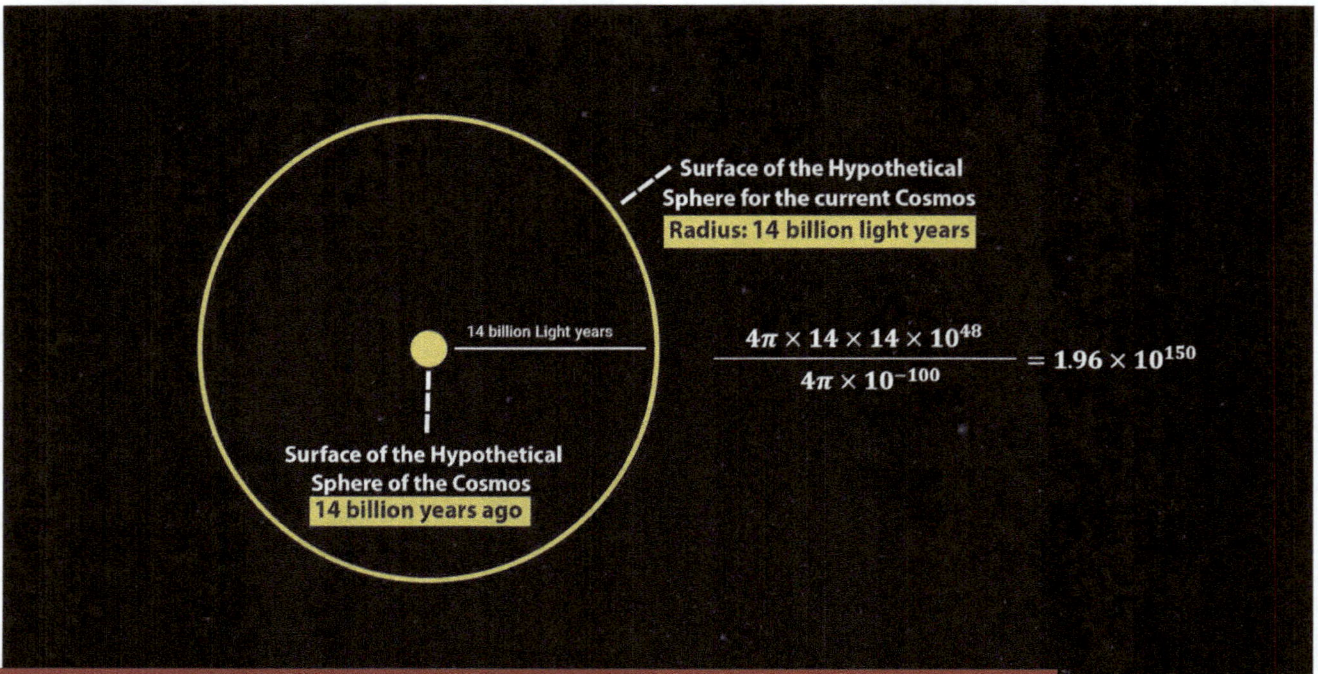

Figure 15: The magnification of the future cosmos compared to the early cosmos

Overall, it can be understood that over time, the cosmos magnifies and while expanding, it exposes a larger observable area to the observer. Additionally, in this magnified volume, a higher density of matter is observed. In other words, from the perspective of an observer on Earth, our current cosmos appears magnified compared to ten billion years ago, and it seems to contain a greater volume of densified matter and energy. Ten billion years later, this magnification will further increase, and this volume will again grow (Figure 16). However, from the viewpoint of cosmologists, the amount of matter and energy in the cosmos should remain constant. Therefore, T-Consciousness Cosmology states that this fact itself is a factor that keeps the fate of the cosmos in the big bang model uncertain; thus:

A magnified density, by more than billions of times, is another reason for the inverse manifestations of the cosmos as depicted by the big bang model.

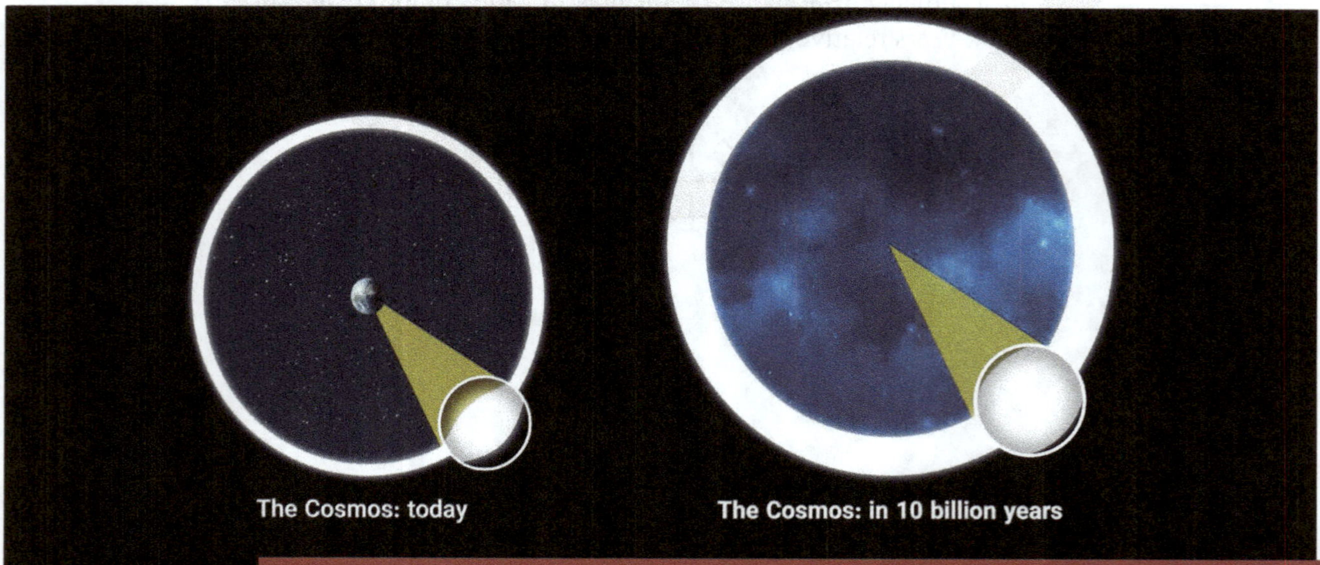

Figure 16: Greater magnification during the expansion of the cosmos

Reason 5:

According to the theory of special relativity, small amounts of mass can be interchangeable with enormous amounts of energy.[20 21] On the other hand, electromagnetic waves present in space also represent a certain amount of energy based on their wavelength.[22] Therefore, as explained in the previous arguments, the increased density of matter in our observable range indicates that the energy emanating from these objects has greatly increased due to the addition of cosmic volume. In fact, with each moment of cosmic expansion, energy generation continues and increases at a very high rate (Figure 17). Consequently, T-Consciousness Cosmology, from another perspective, challenges the big bang model, stating that:

The expanding cosmos has apparently experienced significant energy generation over time, from 13.84 billion years ago, which is another reason for the inverse behavior of the cosmos.

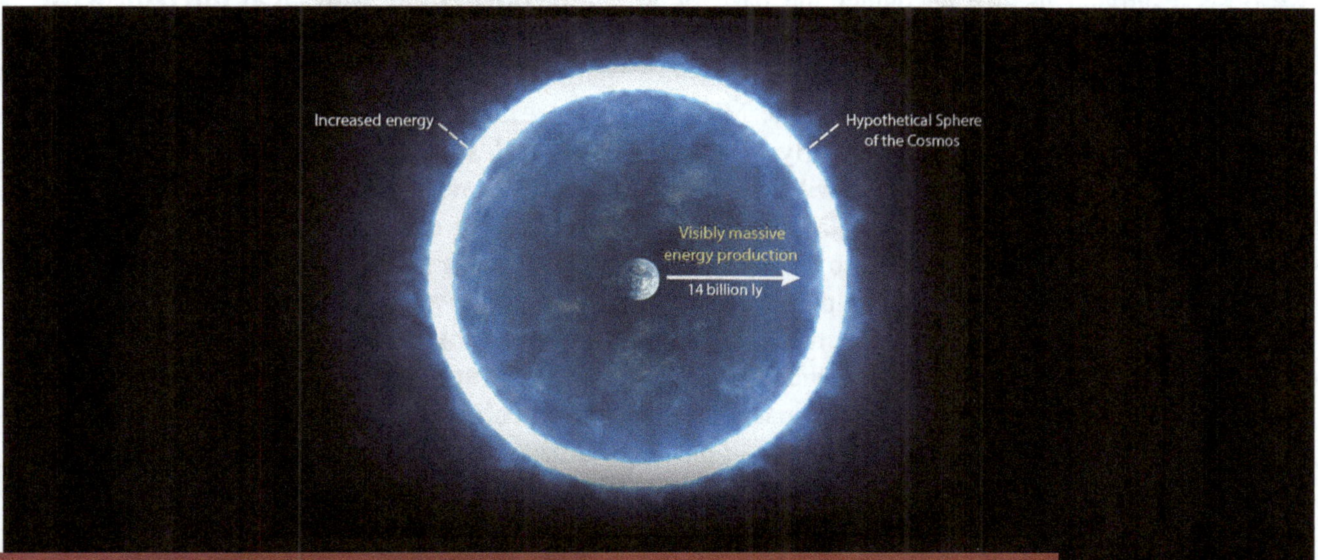

Figure 17: The immense energy generation of the cosmos in the process of expansion

Reason 6:

In line with the fifth reason, if we assume that this energy, as it appears, has been created and expanded over time, it must either be purely potential energy, presenting only a virtual image in our visual horizon, or we are dealing with kinetic energy (Figure 18). In either case, as observers, we are faced with an inverse behavior of the cosmos, and there seems to be no apparent justification for either type of these energies; therefore, from this perspective:

The potential energy, in the visual horizon, or its kinetic state, in both cases, will be another reason for the inverse behavior of the cosmos.

Figure 18
Potential or kinetic nature of energy in the cosmos

Reason 7:

Another inverse behavior arising from the interpretations of the presented models of the cosmos is that, if according to the big bang model, during the expansion of space, we draw a time vector from the hypothetical center of the initial point of the cosmos toward the solar system (our current position), which represents the past of the cosmos until the present time, and continue this vector from the solar system toward the future of the cosmos's expansion, we should face conditions that foretell the future of the cosmos and easily predict its fate. However, wherever we observe around us, we encounter the past, or 13.84 billion years ago. T-Consciousness Cosmology presents its challenge in this way: if, for example, we as observers travel to a galaxy between the Milky Way and the young galaxies of the early cosmos and make observations from there, we will see older galaxies on one side and younger galaxies on the other side. This fact poses a challenge to the big bang model, which suggests that observing the depths of space shows us the past of the cosmos. It appears as if we are in the center of a sphere surrounded by the past of the cosmos, and seemingly in the space-time axis, at this current position, the furthest point of our view with a 360-degree radius is the initial traces of the big bang at a distance of 13.84 billion light-years. While conventional cosmology addresses this challenge with the principle of homogeneity[23], but T-Consciousness Cosmology states:

If we look at the cosmos from the perspective of the big bang model, the direction of the future-facing time vector currently emanating from us in any direction, is hypothetically merged with the direction of the past-facing time vector that surrounds us, and the actual direction of the cosmos's expansion is lost at a certain distance away from us. This leads to a hypothetical spherical horizon with a radius of less than 13.84 billion light-years, inevitably resulting in the existence of a horizon called the "*Timelessness Horizon*" (Figure 19). In fact, within this radius, it will no longer be clear whether our position in the solar system during expansion is moving toward the future or the past of the cosmos. Consequently, from the perspective of T-Consciousness Cosmology, the uncertainty of the main direction of expansion is another reason for the inverse manifestations of the cosmos.

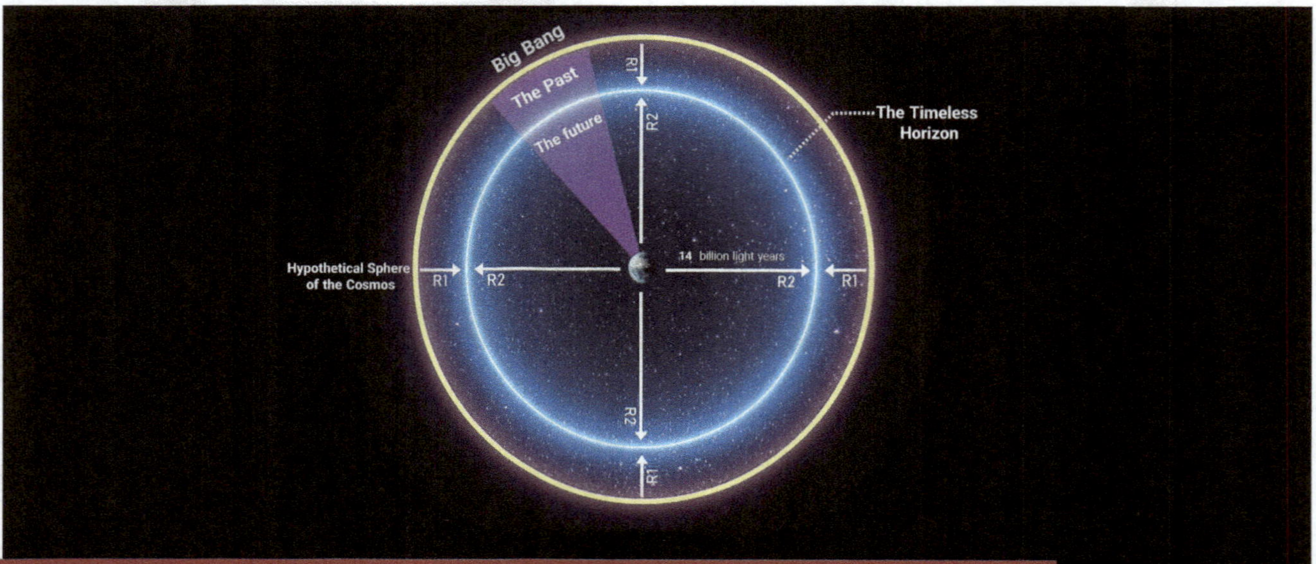

Figure 19: The timelessness horizon from the perspective of T-Consciousness Cosmology in the contexts of the interpretations of the big bang model

Another question that arises is whether, if we could distance ourselves ten billion light-years from Earth, would measurements from all directions still indicate the past of the cosmos at a distance of 13.84 billion light-years (Figure 20)? In other words, will the principle of homogeneity, which has been challenged by recent findings,[24 25 26 27 28] still hold true at that distance?

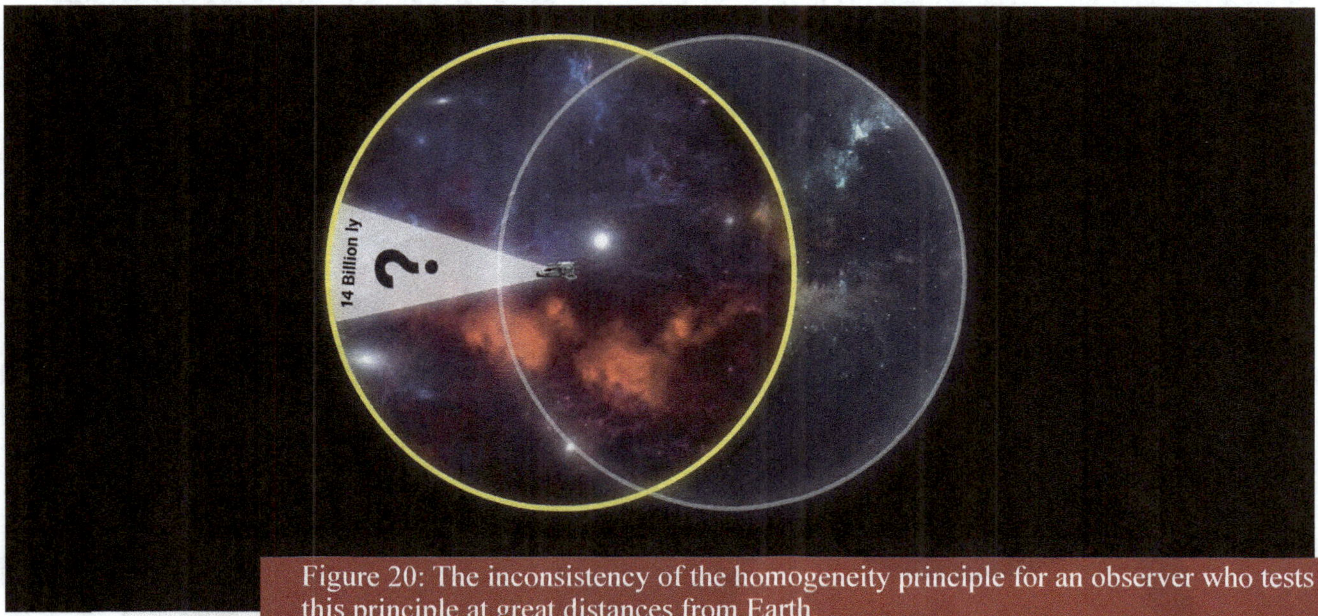

Figure 20: The inconsistency of the homogeneity principle for an observer who tests this principle at great distances from Earth.

Reason 8:

As previously mentioned, according to the big bang theory, the cosmos started expanding from a very small point after the big bang, and it continues to expand like an inflating balloon. In other words, according to this model, the four-dimensional space-time is likened to the surface of an inflating balloon, where the metric distance between celestial objects on this surface is continuously increasing (Figure 21). (It is worth noting that the surface of the balloon is two-dimensional, but space-time is four-dimensional and much more complex than this analogy). In other words, considering the current volume of the cosmos, if an observer observes the depths of space, the location of the big bang will not be visible. This is because, from the perspective of cosmologists, the big bang is considered a moment in time, not a location in space.[18 19]

Figure 21: Balloon analogy to explain the expansion of the cosmos

Considering this model, T-Consciousness Cosmology raises the issue that if we assume the cosmos began expanding from a small point and celestial objects created within the cosmos are distributed across larger scales, how is it possible that what is observed at the furthest edges of the visual horizon (as proposed by cosmologists) is not a small point, but a vast cosmos 13.84 billion years old, in which we currently occupy a position that represents 13.84 billion years into the future after the big bang (Figure 22)? On the other hand, how can there be a situation between that very small point and our current position where, when we talk about the past of the cosmos, it is merely a moment in time, not a place in space?[18 19] In essence, T-Consciousness Cosmology challenges the geometry envisioned by the big bang model, stating:

The change in geometry from a very small point to an immense scale, and the absence of a specific spatial point or temporal interpretation for the beginning of the cosmos according to the big bang model, is another reason for the inverse manifestations of the cosmos.

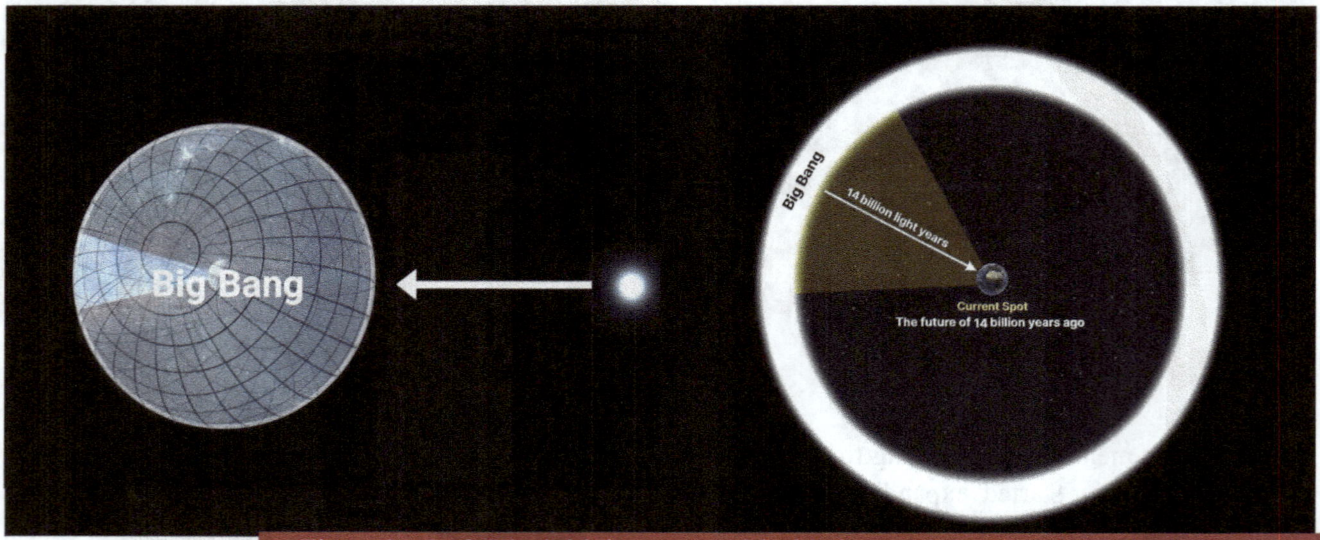

Figure 22: Left: The initial moment of the big bang. Right: The change in the geometry of the cosmos from a very small point to an immense scale

Reason 9:

If we examine the temperature change graph of the cosmos in accordance with the big bang theory, we realize that the temperature at the initial moments of the cosmos was about 10^{27} Kelvin, which gradually decreased with the expansion of the cosmos, reaching 3000 Kelvin after 380,000 years (Figure 23).[29]

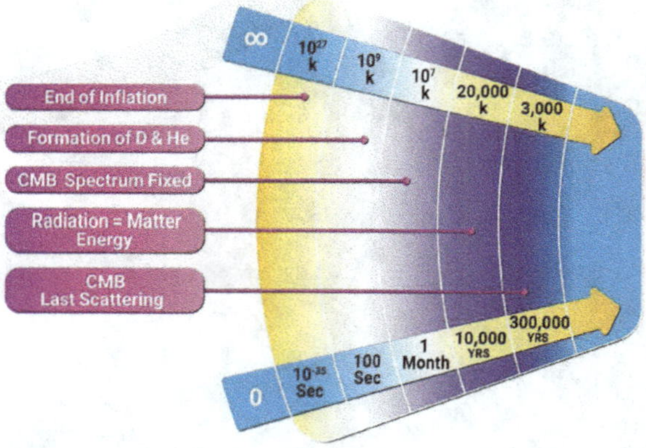

Figure 23: Temperature change chart from the beginning of the big bang to 300,000 years after Its birth

Similarly, as time passes and space expands, the temperature of the cosmos decreases, reaching 2.7 Kelvin today. Therefore, according to the big bang model, when we look at the edge of the visual horizon (13.84 billion years ago), we are actually looking at the past of the cosmos and the initial moments of the big bang when the temperature was at its highest. When we look at the surroundings of the Earth and local galaxies, the temperature has significantly dropped. Based on this, T-Consciousness Cosmology proposes another argument for the inverse behavior of the cosmos, as follows:

If we consider a thermal vector for the cosmos according to the interpretations provided by the big bang model, we realize that, on one hand, during the expansion process, the future-directed thermal vector, or the one extending in radius R2 from Earth in any direction into the depths of space, is headed toward cooler temperatures. On the other hand, the thermal vector emanating from the big bang, which surrounds us in a 360-degree encirclement, is emitted toward us from higher temperatures.

In fact, these two thermal vectors merge at a hypothetical horizon, which T-Consciousness Cosmology calls the hypothetical *"Thermal Obscurity Horizon."* At this hypothetical horizon, the vector indicating the future temperature, reaching the edge of this horizon, has a maximum temperature of 2.7 Kelvin, while the thermal vector we are encircled by, approaching the border of this horizon, is several billion Kelvins (Figure 24).

Consequently, T-Consciousness Cosmology states that if we interpret the behavior of the cosmos according to the big bang model, this hypothetical horizon will be a space where the temperature is subject to uncertainty, and it is unclear what the temperature of this horizon will be. Therefore, the Horizon of Thermal Obscurity is another reason for the inverse manifestations of the cosmos.

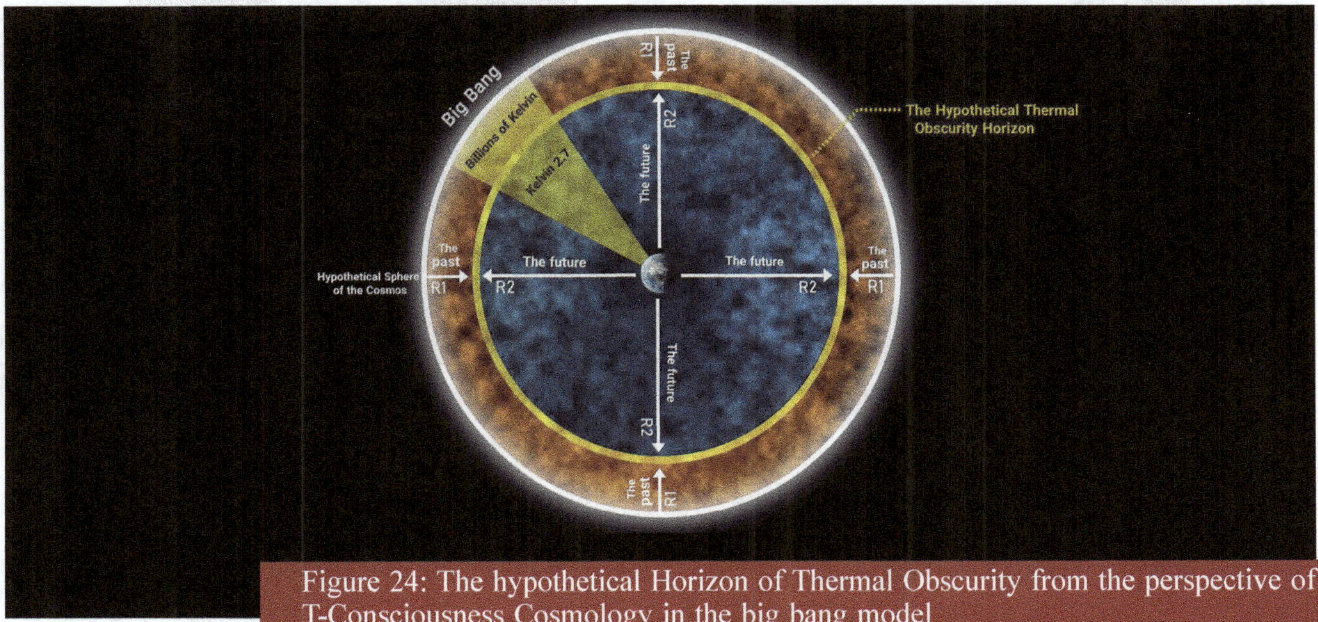

Figure 24: The hypothetical Horizon of Thermal Obscurity from the perspective of T-Consciousness Cosmology in the big bang model

Reason 10:

As stated in reason nine, if we, as observers, want to observe the future direction and expansion of the cosmos along our line of sight according to the fates proposed in conventional cosmology for a flat universe or Big Rip, we should encounter a decrease in temperature from 2.7 Kelvin toward absolute zero in both models, while we will only have a gradual temperature increase in the Big Crunch model. However, contrary to our expectations, at present, based on observations of the farthest distance in any direction, we find ourselves surrounded by temperatures equivalent to several billion Kelvins and this temperature persists despite the cosmos's expansion (Figure 25). From the perspective of T-Consciousness Cosmology, this fact is another reason for the inverse manifestations of the cosmos.

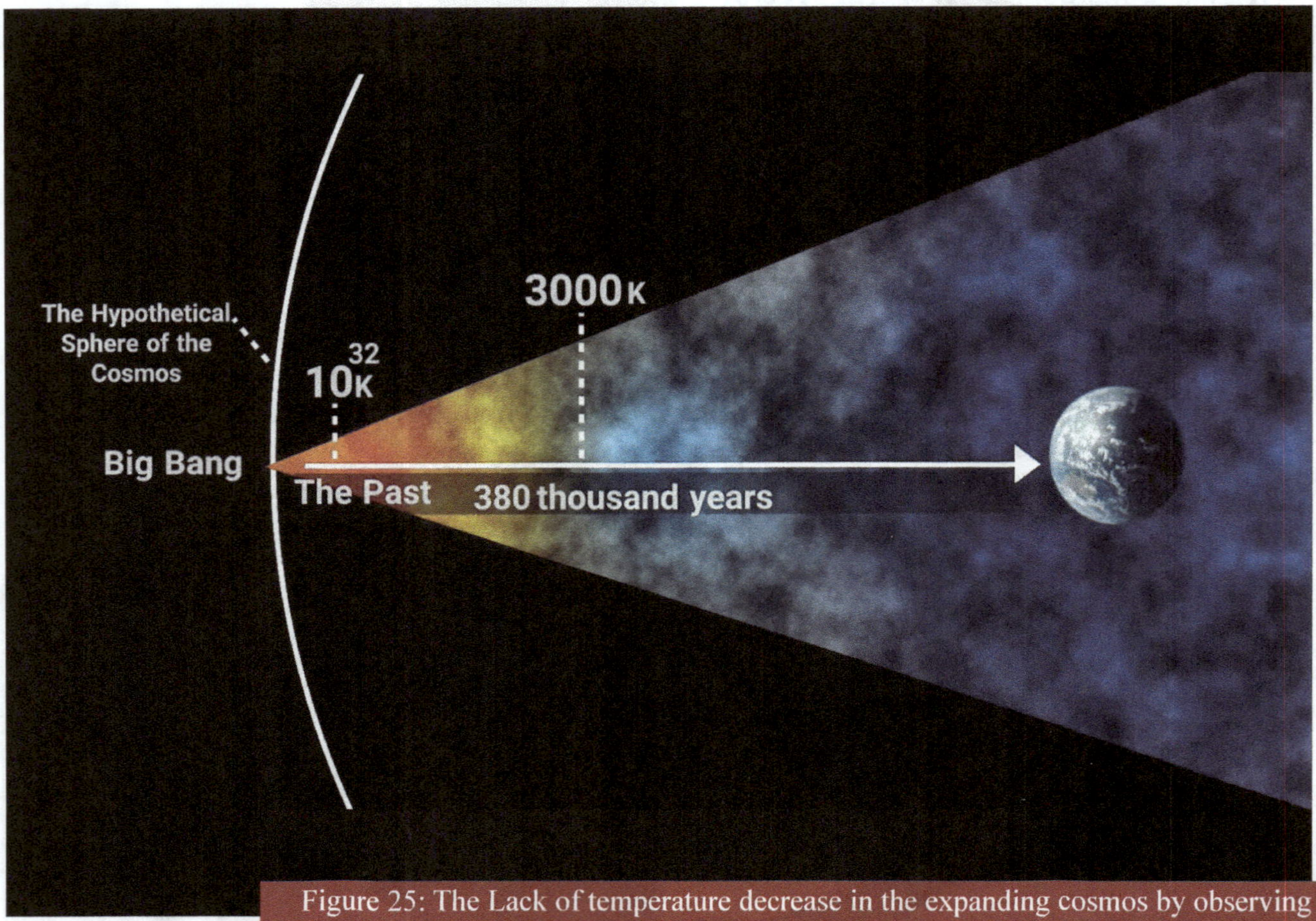

Figure 25: The Lack of temperature decrease in the expanding cosmos by observing the depths of space.

Overall, with the arguments presented under the concept of the 'Inverse Cosmos', it seems that the interpretations of the big bang model in conventional cosmology have led us to observe inverse manifestations in the cosmos, leaving many questions unanswered. Accepting this model requires proposing various mathematical theories, such as the widely accepted theory of inflation, which deviates significantly from the current behavior of the cosmos. Consequently, T-Consciousness Cosmology introduces a new model of the cosmos, called the "*Spherical Cosmos*," proposing hypotheses that can be examined by cosmologists and addresses the challenges regarding the observed behavior of the cosmos.

References

[1] Durrer, R. (2020). *The Cosmic Microwave Background* (2nd ed.). Cambridge: Cambridge University Press.

[2] Saadeh, D., Feeney, S. M., Pontzen, A., Peiris, H. V., & McEwen, J. D. (2016). How isotropic is the Universe? *Physical Review Letters, 117*(13), 131302.

[3] Clifton, T., Clarkson, C., & Bull, P. (2012). Isotropic Blackbody Cosmic Microwave Background Radiation as Evidence for a Homogeneous Universe. *Physical Review Letters, 109*(5), 051303.

[4] Vavryčuk, V. (2022). Cosmological Redshift and Cosmic Time Dilation in the FLRW Metric. *Frontiers in Physics, 10*, 826188.

[5] Bunn, E. F. & Hogg, D. W. (2009). The Kinematic Origin of the Cosmological Redshift. *American Journal of Physics*, 77, 688-694.

[6] Lineweaver, C. H., & Davis, T. M. (2005). Misconceptions about the Big Bang. *Scientific American, 292*(3), 36-45.

[7] Arp, H. C. (1998). *Seeing Red: Redshifts, Cosmology and Academic Science*. Montreal: Apeiron.

[8] Pasachoff, J. M., & Filippenko, A. (2019). *The Cosmos: Astronomy in the New Millennium*(5thed.). Cambridge: Cambridge University Press.

[9] Lewis, G.F. (2016). On The Relativity of Redshifts: Does Space Really "Expand"? *arXiv*, arXiv:1605.08634.

[10] Achúcarro, A., Biagetti, M., Braglia, M., Cabass, G., Caldwell, R., Castorina, E., Chen, X., Coulton, W., Flauger, R., Fumagalli, J., Ivanov, M.M., Lee, H., Maleknejad, A., Meerburg, P.D., Moradinezhad Dizgah, A., Palma, G.A., Pimentel, G.L., Renaux-Petel, S., Wallisch, B., Wandelt, B.D., Witkowski, L.T., & Wu, W.L.K. (2022). Inflation: Theory and Observations. *arXiv*, arXiv:2203.08128.

[11] Cheng, T. (2015). *Inflation and the Accelerating Universe*. Oxford: Oxford Academy.

[12] Peacock, J.A. (2008). A Diatribe on Expanding Space. *arXiv*, arXiv:0809.4573.

[13] Freedman, W. L., & Guth, A. H. (Eds.). (2004). Inflation. In *Carnegie Observatories Astrophysics Series*, Vol. 2, *Measuring and Modeling the Universe*. Cambridge: Cambridge University Press.

[14] Brian Albert, R. (2019). Introductory Chapter: Standard Model of Cosmology. In R. Brian Albert (Ed.), *Redefining Standard Model Cosmology* (pp. Ch. 1). Rijeka: IntechOpen.

[15] Eicher, D. J. (2015). *The New Cosmos: Answering Astronomy's Big Questions*. Cambridge: Cambridge University Press.

[16] Scott, D. (2018). The Standard Model of Cosmology: A Skeptic's Guide. *arXiv*, arXiv:1804.01318.

[17] Li, X., Wang, S., Huang, Q., Zhang, X., & Li, M. (2012). Dark Energy and Fate of the Universe. *arXiv*, arXiv:1202.4060.

[18] Coble, K., McLin, K., & Cominsky, L. (2013). Big Ideas in Cosmology. *arXiv*, arXiv:1303.1768.

[19] Siegel, E. (2023, October 27). *How did the Universe truly begin?* Big Think. https://bigthink.com/starts-with-a-bang/how-universe-truly-begin/

[20] Rahaman, F. (2013). *The Special Theory of Relativity: A Mathematical Approach*. SpringerLink.

[21] Takeuchi, T. (2010). *An Illustrated Guide to Relativity*. Cambridge: Cambridge University Press.

[22] OpenStax. (December 11, 2023). Energy and Power of a Wave. In *University Physics - Mechanics, Sound, Oscillations, and Waves*. Physics LibreTexts.

[23] Peacock, J. (1998). The Isotropic Universe. In *Cosmological Physics* (pp. 65-100). Cambridge: Cambridge University Press.

[24] Jones, J., Copi, C. J., Starkman, G. D., & Akrami, Y. (2023). The Universe is not statistically isotropic. *arXiv*, arXiv:2310.12859.

[25] Ó Colgáin, E. (November 22, 2023). *What is the shape of the universe?* Silicon Republic. https://www.siliconrepublic.com/innovation/what-is-the-shape-of-universe-eoin-o-colgain-study-isotropic

[26] News Staff. (2020, April 9). *Universe is Anisotropic on Large Scales, New Study Suggests*. News Staff. https://www.sci.news/astronomy/anisotropic-universe-08312.html

[27] NASA. (2020, April 7). *Universe's Expansion May Not Be The Same In All Directions*. NASA. https://www.nasa.gov/universe/universes-expansion-may-not-be-the-same-in-all-directions/

[28] Koberlein, B. (2020, April 9). *New observations show that the Universe might not be expanding at the same rate in all directions*. Universe Today. https://www.universetoday.com/145605/new-observations-show-that-the-universe-might-not-be-expanding-at-the-same-rate-in-all-directions/

[29] Coble, K., McLin, K., & Cominsky, L. (n.d.). The Friedmann Equation and the Fate of the Universe. In *Big Ideas in Cosmology*. San Francisco State University, Chico State University, & Sonoma State University. Physics LibreTexts.

www.ingramcontent.com/pod-product-compliance
Lightning Source LLC
LaVergne TN
LVHW062256071126
851857LV00050B/1852